ODD CORNERS
OF THE
EASTERN

ODD CORNERS
OF THE
EASTERN

ERIC SAWFORD

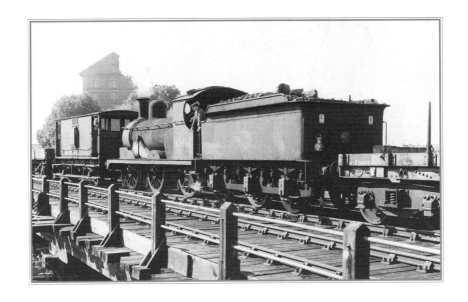

First published in 2001
This edition published in 2010

The History Press
The Mill, Brimscombe Port
Stroud, Gloucestershire, GL5 2QG
www.thehistorypress.co.uk

British Library Cataloguing in Publication Data.
A catalogue record for this book is available from the British Library.

ISBN 978 0 7524 5808 3

Typesetting and origination by The History Press
Printed in Great Britain
Manufacturing managed by Jellyfish Print Solutions Ltd

CONTENTS

Living for many years in Huntingdon, I was able to take many photographs of the two stations there, as well as various other railway installations, over the years. This picture of Huntingdon East shows the small sub-shed in the distance, and provides a wealth of detail, ranging from the lattice-type overbridge, the gas lighting and the water-crane to check rails on the track and signals. (25.12.55.)

B1 no. 61047 runs into Cambridge with a train from Norwich. On the left is part of the motive power depot yard, with an E4 2–4–0 standing ready to leave for carriage pilot duties. Note the number of tracks to the right of the picture; these included the goods-avoiding lines. (22.9.53.)

INTRODUCTION

Fortunately, on many occasions I took the opportunity to record on film some of the numerous interesting items of equipment and scenes that existed on the Eastern Region in steam days. Although my principal interest was steam locomotives, the railway infrastructure that existed in the 1950s and 1960s was very different from that of today. Rail journeys offered many attractions, and wayside stations still had their goods sheds and sidings. On the East Coast main line there were special water-troughs to enable express locomotives to replenish their tenders at speed. The adjacent water-softening plants in turn used old locomotive tenders to carry away the resultant sludge. These are just a few examples; numerous others will be mentioned later.

For this book I have chosen many different types of locomotive, with a variety of scenes and items, giving as wide a range as possible. One section that always attracted my attention was the Engineering Department. Unlike today, where contractors carry out the lineside work, in those days it was all in the hands of British Rail staff and the equipment available was very different. Rail was laid in sections and it was by no means uncommon to see a large gang of men slewing a section of track into its final position. Always of interest were the engineering trains themselves, many of which incorporated in their make-up old coaches, some dating back to well before Grouping. On the Eastern Region the engineering trains included numerous veteran six-wheelers in sombre plain black livery. Machines did exist to carry out ballast tamping, although much of this work was still done by hand. Old coaches also served in numerous other roles, such as mess rooms or stores, or with the signal and telegraph department. Occasionally one might be pressed into service for specialist work, for example as a weighing machine testing unit.

In those days you could not travel far before you passed a signal-box. These varied considerably in size and design, but each one controlled the semaphore signals and points on its relevant section. Signal wires, telegraph poles and point rodding were all lineside features. The signals themselves ranged from the large gantries found at principal rail centres to single posts, plus the ground signals (often referred to as 'Dollies') controlling points. In time these were all to be replaced on many routes by colour lights and point motors, controlled from a power box many miles away.

Station equipment would usually include one or more water cranes so that steam locomotives could replenish their supplies. The cranes were very vulnerable to frost damage, and to prevent this each had its attendant coke- or coal-fired heater with a flue and twin outlets to keep the supply pipe free of ice. The heaters, of course, required frequent attention to ensure that the fire did not go out, but nevertheless in severe weather freezing could – and did – occur. Heavy four-wheeled trolleys were another station feature; in those days parcel trains often delivered large bulky items, especially at larger stations where both parcel and letter mail was received and dispatched.

The goods shed was another common feature. In the adjacent yards you would invariably find an unloading dock, sometimes with an ageing crane that saw little use. In the yards a loading gauge was used to check bulky and difficult consignments so that they were not released 'out of gauge'. Agricultural machinery could often be seen loaded on flat wagons. Even small wayside stations would invariably have some wagons present, such was the amount of traffic handled by the railways. The 'pick up goods' was a common sight, its job to collect and deliver wagons for onward movement to large rail centres, where they would be marshalled into principal goods services to continue their journey.

There was a whole host of different wagon types. Vans were commonplace; some were in general use while others were retained for special loads, such as fruit, fish and even gunpowder. Coal traffic from the extensive Midlands and Yorkshire coalfields resulted in numerous coal trains carrying supplies for commercial, domestic and, of course, locomotive use (not to mention the returning empties). In the early 1950s coal trains were a very familiar sight on the East Coast main line, usually hauled by ex-WD 2–8–0s, known as 'Austerities' but referred to by enthusiasts as 'Dub Dees'. With the introduction of the Standard locomotive designs the 9F 2–10–0s soon became involved in this work. Initially they experienced some difficulties with braking problems but these were soon rectified following tests. Wagons used in these trains included many of the steel mineral type built by British Railways' own works. Numerous examples of the wooden-bodied wagons were still very much in evidence at this time, and occasionally old colliery names could just be made out under the accumulated layers of grime.

There were countless types of special wagon to be seen and photographed, especially in the 1950s. Most enthusiasts did not give them a second glance, as it seemed at the time that they would be a feature of our railways for ever. That they could disappear was unthinkable – but how wrong we were. The types to be seen included bogie brick wagons, flat wagons of four-wheeled and bogie types, 'well' wagons for transporting large bulky pieces of machinery, and tank and sand wagons. Small containers were moved by rail on special flat wagons; fully fitted, these flat wagons were capable of running in fast goods trains. Once at their destination, they were loaded on to a British Railways lorry and delivered direct to the customer's premises.

One of the loneliest jobs on the rail system was that of the goods guard, who spent hours in isolation in his brake van at the end of a long train. In those days there was no communication except by hand lamp. On busy routes, or at busy times, trains might be left standing in sidings for a considerable time while a clear path was found, with the goods guard waiting patiently in his van. Each of the four railway companies had contributed its own basic brake van designs at Nationalisation, and examples were often to be seen well away from their usual haunts. Even bogie brake vans were not unknown. One of the most essential items of these vans was the coal-burning stove, not just for heating but also for heating food for the guard, especially on a cold winter's day.

Locomotive depots and engineering yards were good hunting-grounds for those interested in wagons and coaches. The latter were often to be found in the make-up of breakdown cranes employed as mess and tool vehicles. The Engineering Departments also used them in their trains, principally for similar duties. At some depots even older vehicles could be seen, minus wheels and running gear, converted to mess and stores use. Doubtless many of these were eventually to go up in smoke on site. Wagons to be found at engine sheds included vehicles for transporting sand, oil and other essentials; in hard water areas old tenders were used to collect the lime extracted from the water supply. Some still carried their tender plates and, in many cases, railway company lettering could be seen. Old tenders were also used for other duties, for example as snowploughs and even as water carriers.

Breakdown trains were called out at short notice to deal with any emergency, and were manned by fitters under the supervision of senior depot staff. These units were often kept in light steam, which enabled the crane to reach full working order by the time it arrived at the scene. The steam-powered cranes varied in size, from the large units at major depots to the smaller (and usually older) ones elsewhere. When a call

Unfortunately I was never able to visit the Wisbech & Upwell Tramway when these tram engines were in charge. This picture, taken at March depot, shows four J70 class engines and a solitary Y6 in store, their duties having been taken over by a diesel. The J70s were transferred away and soldiered on for a short time but the solitary Y6 was not so lucky: it was withdrawn in the same month this picture was taken. (8.11.52.)

was received the first available locomotive was commandeered, which was fine if it was capable of reasonable speed. I have on more than one occasion witnessed the 45-ton New England crane headed by a WD 2–8–0 travelling flat out on the main line.

The East Coast main line was very important for mail services and lineside apparatus for the TPOs (Travelling Post Office) could be seen at various locations. Huntingdon, for example, had installations on both the Up and Down lines. The first train of the night was the London–York–Edinburgh, due at Huntingdon shortly before 10 p.m.; this train both dispatched and collected pouches. Depending on loading, two sets were usually used for dispatch; this involved four pouches each containing one or two mailbags. At each site there was just one net. As the train roared through Huntingdon, often with an A4 in charge, its dispatcher would look for the sighting-board alongside the main line, which was the signal to swing out the arms from which the pouches were suspended. The train was capable of dispatching several pouches, and as the bar on the net hit the pouch arm they slammed hard into the receiving net. Just a few feet beyond the net was the first of the Down side dispatching sets, each consisting of two arms, enabling two pouches to be dispatched. In all, four sets existed on this side, equally spaced a considerable distance apart.

The method of operation was for two men to take the outward mail to the lineside and prepare the heavy leather pouches. A length of soft brown string would be threaded through the pouch so that each end could be tied to the platform eyelets to prevent the pouches swinging about as the train passed. Everything could be made ready but the mail

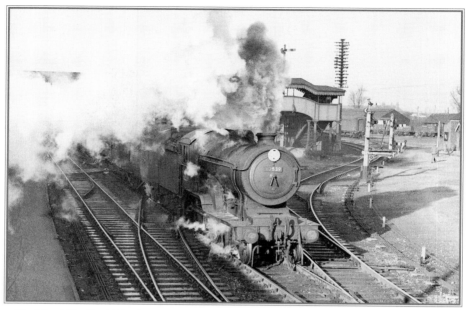

The 'Claud Hamilton' D16 class 4–4–0s were the mainstay of passenger services in East Anglia for many years, but unfortunately no example has survived into preservation. Here, no. 62539 blasts away from March with a train for King's Lynn. The 'Clauds' were express engines in their heyday and were capable of a fair turn of speed. (8.11.52.)

for dispatch could not be swung out and tied until the next train was the 'mail'. Contact with Huntingdon signal-box was by a rather primitive telephone.

Working on the TPOs was fine on a summer's evening, but it was all very different when the weather was frosty, snowy or foggy and the train was running late. Each dispatching set had its own hut, but there was no form of heating. An additional hazard on the Down side was that the operators had to cross the slow line to get to the TPO equipment. Care was certainly needed in foggy conditions. Normally everything worked well, but there were odd times when there were problems, usually caused by a pouch swinging about on the train, hitting the bar wrongly and dropping down the embankment.

The second mail train on the Down side was the North Eastern TPO at around 11.30 p.m. Mail arrived by road from Cambridge for this train and on occasions all four sets could be used. The last train of the night was the North Eastern TPO Up at around 2.30 a.m., which delivered mail for Huntingdon and Cambridge, and collected late-posted items for first delivery in London. The TPO coaches carried a full complement of sorters and staff for operating the dispatching and receiving equipment. In addition, letters could be posted in a box on the train. TPOs still run, but the lineside equipment has long since disappeared. Inspectors were frequent visitors to these installations, and operating staff were passed to work them. The nets, which were made from heavy rope, took a continuous battering from the pouches and needed to be changed on a regular basis.

During the 1950s numerous branch lines were still in operation, usually with ageing tank locomotives. It was a decade of change and many were to close within the next few years. Some retained goods services for a short time before these too were withdrawn. By the 1960s the rail system of the Eastern Region was very different. Most of the closed

All traces of the condensing gear once fitted to N1 class 0–6–2T no. 69443 had long since gone when this picture was taken at Bradford motive power depot. N1s were used on local passenger services in the West Riding for a number of years. Many were once a familiar sight in the London area, working on carriage pilot duties and inter-regional freight. (24.6.56.)

lines had already been lifted and nearly all traces of the existence of a railway had gone; nature soon reclaimed the tracks and the once neat and tidy embankments and cuttings.

Faced with the imminent closure of branches and cross-country lines, the leading railway societies organised numerous rail trips. Many incorporated some branches which were goods only lines, among them the Benwick branch, with passengers travelling in open wagons. Some of the branches visited had long since lost regular passenger services. The former Great Northern & Great Eastern Joint line from Somersham to Ramsey East was a prime example, with regular services having been withdrawn on 22 September 1930. It had remained open for goods traffic, and during the summer excursions were operated to Yarmouth Races, normally worked by a 'Claud Hamilton' D16 class 4–4–0. A special organised by the Railway Correspondence & Travel Society included this branch in its itinerary, worked by J17 no. 65562 of March depot, in immaculate condition, and carrying a large headboard for 'The Fensman'.

At one time Ramsey had two branches providing services. The Great Northern Railway line from Holme on the East Coast Main line ran across the Fens to Ramsey North. The passenger services on this branch were withdrawn on 6 October 1947, but here again the line remained open for goods traffic for a number of years.

Another branch that attracted a considerable number of enthusiasts in its final years was Mildenhall, not because of the scenery or any gradient involved but principally because of the motive power, which was a Cambridge depot E4 2–4–0 in the early 1950s and a J15 0–6–0 in later years. (If you were unlucky you might have found an Ivatt 2MT 2–6–0 in charge.) It was a typical branch line. On arrival, the locomotive used the small turntable alongside the signal-box and then rejoined its train. This gave the enginemen a bit of a break. Passenger numbers were low and on 18 June 1962 this service was withdrawn. But it was not only the branches that fell victim to closure: the M&GN and cross-country lines were soon to join the growing list of withdrawn services.

During the summer months many lines handled excursion trains, especially at weekends. These were usually full, and the seaside resorts that were their destination received a deluge of visitors. From 11 a.m. onwards everything was busy until the trains departed from late afternoon onwards. These excursions often resulted in locomotive types from other regions working in, attracting the attention of enthusiasts en route. London Midland Region class 5s often appeared at Yarmouth, for example. In addition, the annual summer two-week shut-down of some towns generated extra traffic. One trip that used the East Coast main line for a number of years ran from King's Cross to Skegness; worked by a King's Cross B1 class 4–6–0 throughout, it picked up at several stations. As the numbers of private cars rapidly increased, the demand for excursions gradually dwindled. Now long gone is the once-familiar sight of locomotives and trains serviced and ready for the home run at Skegness, Hunstanton and other East Coast resorts. In some cases it's not just the trains that have gone – so has the railway itself.

During steam days it was the named locomotives that attracted the most attention. Enthusiasts referred to special workings and sightings by the engine's name, hardly ever by its number. The Eastern Region locomotives had a wide and fascinating selection of names: birds, types of antelope, racehorses and individuals with a railway background. There were also some that did not fit into any category, *City of London* and *Royal Sovereign* among them. Examples of some of the nameplates are included in the photographic section.

For several years I held a Lineside Photographic Permit, which enabled me to record on film some of the once-commonplace railway features. I must admit that at the time it seemed impossible that they would not remain in situ for ever. As will be seen, a number of these pictures were taken around my home town, Huntingdon, with its two stations, and in the process I built up an extensive collection showing all aspects of the railway which existed at the time. Nowadays these pictures are of considerable interest to railway modellers keen to ensure that their work is as authentic as possible. Also included are photographs of a very small number of diesels, early types which have themselves become part of railway history. By the end of the 1950s it was already becoming apparent that the steam locomotive's days were numbered, despite the fact that steam engines were still being built. Yet many enthusiasts still paid little attention to the new diesels (apart from those interested in numbers).

As has already been well documented, some of the diesel designs remained in service for only a comparatively short time, and with hindsight I wish I had recorded more on film. The 1960s were to be depressing years for steam lovers; as the numbers of diesels continued to rise, increasing numbers of steam locomotives were replaced, put in store or simply cast aside. It was not long before examples of the Standard designs were among these outcasts, their years in service counted in single figures. In the headlong rush for progress, the less troublesome and more comfortable diesels attracted many enginemen, with steam being regarded by many as outdated and old-fashioned. Maintenance standards quickly fell; indeed, many locomotives were in a deplorable condition, covered in grime and dirt. The programme of closure of steam depots on the Eastern Region commenced and continued at an ever-increasing rate. Before long famous sheds, such as King's Cross 'Top Shed' (June 1963), Stratford (September 1962), March (November 1963) and New England (January 1965), had all gone – in less than two years. This gives some indication of just how quickly dieselisation took place. The last depots to operate steam were in the northern part of the region.

Chapter One

STATIONS AND LAYOUTS

Today's railway scene is, of course, very different from that of steam days. Numerous stations have been rebuilt, and countless branches and small stations have closed, often leaving no trace to show that a railway once existed there. Even small wayside stations usually had a goods shed and sidings, as well as a cattle dock and a loading stage for bulky items. Farm machinery was frequently transported by rail in those days. Loading gauges were a familiar sight, ensuring that wagons were not sent out with loads that would not clear platforms, bridges or tunnels.

Country stations were often situated at a considerable distance from the village they served. Stopping passenger trains on the East Coast main line were frequently referred to as the 'Parley'. With the number of passengers rapidly declining, many of these stations soon closed in British Railways days. Even small ones had a stationmaster and porters, together with other staff to look after the rapidly decreasing amount of goods traffic.

Rail layouts were much more complex in those days, but the surplus tracks and numerous sidings have long since gone, as has much of the lineside equipment which was once a common feature of our railways – notably the countless thousands of telegraph poles and associated wiring. A few platelayers' huts still exist, mainly on secondary lines, but most are now disused.

Some of the following pictures contain a wealth of information for railway historians and of course for modellers keen to include as much detail as possible on their layouts.

One event that always attracted the attention of enthusiasts at King's Cross during the summer months was the arrival of 'The Elizabethan'. This train was worked by Haymarket (Edinburgh) or King's Cross A4 Pacifics. Here no. 60009 *Union of South Africa* backs out of the station for servicing after the long non-stop run from Edinburgh. (9.61.)

The driver of N2 no. 69560, a non-condensing example of the class, and the wheel-tapper deep in conversation while the guard chats to the locomotive fireman. N2 0–6–02Ts were a familiar sight at King's Cross for a great many years, both on suburban duties and, as in this case, on empty stock workings. (9.7.53.)

This is a picture that will be familiar to anyone who can recall King's Cross in steam days, with its complex track layout. In the background A4 no. 60032 *Gannet* is about to enter Gasworks tunnel. The locomotive servicing point, part of which is just visible on the left-hand side, was handling both steam and diesel at the time. (9.61.)

Having been released from its train, A4 no. 60034 *Lord Farringdon* backs out of the station. This engine was allocated to King's Cross depot for many years. Note the signal indicating M1. (9.61.)

The massive water-crane at the north end of Cambridge station was frequently used during the day. Here, E4 no. 62781 on station pilot duty takes the opportunity to replenish its tender. This particular E4 was one of the batch fitted with a side window cab for duties in the north-east. Note also the impressive array of signals, part of which can be seen in this picture. (10.10.55.)

Hitchin station was a popular vantage point for enthusiasts in steam days, providing access to both the East Coast main line and the Cambridge line. In addition, the Bedford line, until closure, brought LMR locomotives to Hitchin. The Cambridge trains to King's Cross were worked by B1s or B17 Sandringham class 4–6–0s. Here, no. 61653 *Huddersfield Town* runs into the station with a six-coach train. This engine was built as a B17/4 in April 1936, rebuilt B17/6 in May 1954 and withdrawn in January 1960. (14.10.56.)

This picture is sure to bring back many pleasant memories to those who can recall the north end of Cambridge station, with its impressive array of signals, during the 1950s. On the left can be seen part of the busy yard of Cambridge motive power depot, widely known to enthusiasts of the day as 'the dump'. (17.5.52.)

During the early and mid-1950s Cambridge was one place where you stood a good chance of seeing 2–4–0 tender locomotives in action, with most of the remaining E4 class allocated there. They were often used for station pilot duties, on which no. 62796, shown here, was employed in the late afternoon sunshine. (10.10.55.)

Brightlingsea station, with its typical platform roof. Note the fancy cast-ironwork at the top of the end pillar and the barrow, similar to those found at most stations. Lighting was by gas lamp. The water-crane and heater still remain intact at the end of the platform. (7.9.62.)

Mildenhall was a typical branch line terminus. J15 no. 65451 stands ready to leave for Cambridge with the late afternoon train comprising just two coaches and a van. J15s were the normal branch motive power but E4 2–4–0s and Ivatt 2–6–0s were not unknown. (31.5.56.)

On arrival, the Mildenhall branch locomotive was turned and quickly rejoined the train. The engineman then had time for a break before the return working to Cambridge. Here the guard lends a hand to turn J15 no. 65475. (2.7.55.)

As was the case with many East Anglian branches, the Mildenhall service was not particularly well patronised. Here, J15 no. 65475 stands at the head of the afternoon two-coach train to Cambridge. (2.7.55.)

The driver of J15 no. 65451 climbs aboard after turning at Mildenhall. The services on this branch were mostly worked by one of the small stud of J15s allocated to Cambridge, with E4 2–4–0s and Ivatt 2MT 2–6–0s appearing at times. (31.5.56.)

The track layout to the north of Sandy station, photographed in the mid-1950s. To the right of the picture is the Cambridge–Bedford–Bletchley line that climbed up and over the main line. (31.7.54.)

Sandy station from the south. The East Coast main line is on the left. The signal-box and station to the right served the Cambridge–Bletchley line, and this closed many years ago. Sandy has since been remodelled. Note the cattle wagons on the left of the picture. (29.4.56.)

Working a light engine north on its journey to Doncaster Works was an easy run for London-based enginemen. There was a fairly regular path following the first northbound semi-fast passenger service. Here, N2 no. 69556 is heading north on the main line. (17.11.51.)

The Kettering–Cambridge afternoon train awaits the right away from St Ives to continue its journey to Cambridge. This service ran over Eastern rails from Huntingdon East. After closure the line was still used for sand trains to Fendrayton. All traces of the railway have long since gone from St Ives but for many years now there have been discussions about the possible reopening of the line to relieve pressure on the heavily congested A14. (8.10.51.)

St Ives, photographed from the end of the platform looking towards Cambridge. The two tracks on the left were the March line, known as the 'St Ives loop'; those to the right were the Kettering line via Huntingdon East. Nothing remains to indicate that a railway ever existed here. (17.3.54.)

Considerable civil engineering work – as shown here – was required to add an Up slow line from Abbots Ripton to Huntingdon. Despite all this work the line was later removed, although there is a chance it may now be reinstated. (20.9.59.)

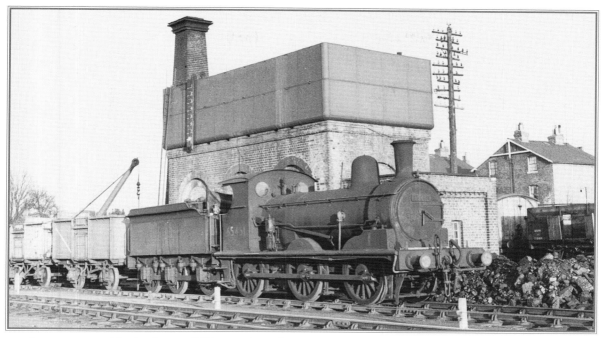

It was most unusual for the pilot, no. 65451, to stand near the water tank at Huntingdon instead of at its usual sub-shed at nearby Huntingdon East. Presumably this was because of the Christmas holiday. (25.12.55.)

Ramsey East station during the 1950s was only used by 'Race Specials' to Yarmouth during the summer. Regular passenger services on the branch to Somersham ceased on 22 September 1930. (24.7.55.)

The immaculate J17 class 0–6–0 no. 65562 heads the six-coach RCTS 'Fensman' towards Ramsey East. Enthusiasts' specials were organised to include various branches at this time, including the freight-only Berwick branch – where participants left the luxury of the coaches to travel the branch in open wagons. (24.7.55.)

Ramsey East station had long since lost its regular passenger service when this picture was taken. It did, however, see occasional specials to Yarmouth races and, as in this case, an enthusiasts' special organised by the Railway Correspondence & Travel Society. 'The Fensman' is pictured ready to return to Somersham, connecting with the March–Cambridge line. (24.7.55.)

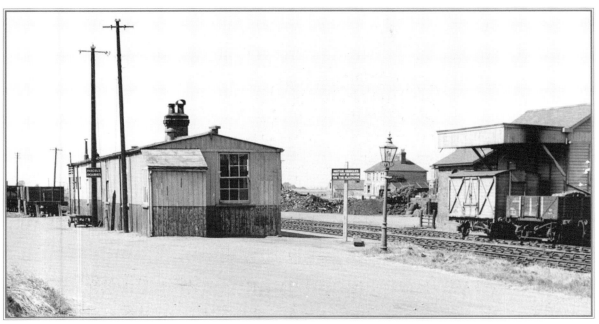

The station buildings at Ramsey North after passenger services had ceased. The branch was still open for goods traffic. Note the small goods shed, gas lighting and sign reading 'Motor vehicles must not be driven on the platform'. (24.7.55.)

Ramsey boasted two stations, both of which had lost their passenger services when this picture was taken. This is another shot of Ramsey North, which connected with the East Coast main line at Holme. The branch lost its passenger services in 1947. (24.7.55.)

Huntingdon East consisted of three platforms, only one of which (out of picture to the right) was regularly used by passenger trains on the Kettering line. Here, L1 class 2–6–4T no. 67745 is heading the Fridays only Royal Air Force leave train, comprising a close-coupled suburban set. (9.5.54.)

Grass and weeds had already started to grow extensively around the redundant locomotive shed at Huntingdon East when this picture was taken. The shed itself was being used to store a set of crossing gates. Within a short time the building was reduced to rubble. (31.8.61.)

These two platforms at Huntingdon East were not regularly used after the early 1950s when the Huntingdon–St Ives service was withdrawn. To the left of the island platform was the 'Midland' platform used by the Kettering–Cambridge line until closure in 1959. The tracks in this picture had connections to the East Coast main line. (25.12.55.)

Although the platform was used only by London Midland Region Kettering–Cambridge line trains it has been included as it shows the relationship to the East Coast main line. In the distance is the Huntingdon no. 1 signal-box that also controlled the East station workings. Note the check rail and severe curve. (6/54.)

Huntingdon East in the mid-1950s. To the left are the two platforms once used by the local Huntingdon–St Ives service. In the foreground is the 'Midland' platform with its severe curve, check rails and 10mph speed restriction. (6/54.)

Huntingdon East presented a strange sight when this picture was taken. The island buildings had completely gone but the main station still remained intact. The water-crane at the end of the platform was still being used by the J15 0–6–0 on track recovery. The site of this station is now part of the main line station car park. (31.8.61.)

The London Midland Region weekday goods from Kettering taking on water at Huntingdon East station before running the 2 miles to Godmanchester, its destination. Johnson 2F no. 58193 was one of a batch allocated to Kettering. This service frequently arrived with just a brake van at this time, often returning with the same load. The main traffic, such as it was, was limited to stations nearer to its home base. Note the repainted chimney and smokebox often seen on LMR engines after light repairs. (17.6.52.)

Godmanchester station with the St Ives–Huntingdon local goods ready to complete its return journey. In the 1950s the service acted as an interchange, loads to and from the East transferring at Huntingdon to the Eastern Region main line. Anything to and from the LMR transferred at Godmanchester was handled by the weekday Kettering–Godmanchester pick up goods. Note the high-sided wagon in this picture. (7.6.52.)

Godmanchester is one of countless stations long since closed, with no trace of the railway remaining. The last passenger services to use this station were the LMR Kettering–Cambridge trains that finished in June 1959. (7.6.52.)

A stranger to the East Coast main line, Britannia class no. 70020 *Mercury* heads a Home Counties Railway Society special to York. It is seen here taking on water at Peterborough. The original request was for a Princess Coronation class locomotive, but none was available and the Britannia was substituted. (4.10.64.)

Several sets of Travelling Post Office (TPO) lineside equipment were to be found at Walton, just north of Peterborough. Here, Ivatt 4MT 2–6–0 no. 43063 heads towards the north station with an M&GN line train. (24.9.55.)

During the early 1950s V2s were often seen on express duties, and these 'maids of all work' handled them in their stride. Here, no. 60975 starts a King's Cross to Edinburgh train from Peterborough North. Numerous once-familiar pieces of lineside equipment can be seen in this picture. (5.9.53.)

A grimy V2 no. 60912 restarts a heavy express from Peterborough North. In those days there was a sharp curve through the station which required the use of a banker to assist the train locomotive. At this time veteran C12 class 4–4–2Ts performed these duties. Note the Great North Railway trespassing notice in front of the water-crane. (5.9.53.)

It took many years to recover from the wartime demands on the railway, especially catching up with the backlog of locomotive maintenance due to shortage of manpower. The V2 2–6–2s, designed by Sir Nigel Gresley, performed numerous herculean tasks during the war and were invaluable right up to the end of steam. No. 60914 is seen here running into Peterborough with a relief King's Cross express. (5.9.53.)

The sharp exhaust note of the ageing C12 4–4–2Ts was a familiar sound under the overall roof which existed at Peterborough North in the 1950s. These locomotives were employed on banking heavy northbound expresses and, as no. 67365 is doing here, on stock workings, in this case marshalling the coaches for a stopping service to King's Cross. (5.9.53.)

Skegness station was a busy place during summer weekends, with normal services and excursions from many places including the Midlands and London. Here, K2 no. 61745 of Boston awaits departure time on a busy Sunday. (19.6.55.)

Huntingdon North, pictured from the south end, with the goods shed and yard on the left. At this time the platform could accommodate no more than six coaches. The station and its layout have changed considerably, and only the station building on the right remains. Long gone are the water tank and sidings. (9.9.54.)

Lincoln St Marks station with Director D11 class 4–4–0 no. 62666 *Zeebrugge* heading a Sunday service to Nottingham and Derby over the former Midland route. The five Lincoln-based D11s were transferred to Sheffield Darnall in 1957. (14.8.55.)

During the mid-1950s I took a number of pictures of Huntingdon North, which certainly presented a very different scene from today's. The station platforms were much shorter then. Locomotives on express duties and goods short of water occasionally used the water-crane that can be seen here. The signal had recently replaced an earlier design. At the time goods traffic required a J15 for shunting and daily trip working to St Ives. (8.4.56.)

Huntingdon North from the south. Again, it is very different today: the island platform and footbridge were demolished and completely rebuilt, lengthened and reduced to just one platform. The main station buildings still remain, with a much longer platform and bay. Passenger numbers using the station today are much greater, with large numbers commuting daily to London. (8.4.56.)

Huntingdon North looking south, showing the island platforms. The station's name changed to just Huntingdon after the closure of the East station. When the station was rebuilt, the footbridge was demolished and a new open-type unit installed in a different position. (8.4.56.)

The road bridge in this picture still carries heavy traffic, but a new concrete bridge was erected above this to carry the very busy A14. The platform on the left was completely demolished, and its replacement relocated to end some yards from the bridge. Colour lights have now replaced the tall signals. (8.4.56.)

The footbridge at Huntingdon was starting to show its age when this picture was taken. Both the island platform and the footbridge were demolished when the station was rebuilt. All that remains now are the far buildings that were incorporated into the considerably lengthened platforms. (6/52.)

Coal was still being widely used for domestic purposes in the 1950s and 1960s. Most stations had one or more coal merchants working from the goods yards, as in this picture taken at Huntingdon. (6/52.)

Water-cranes were to be found at the end of both Up and Down platforms at Huntingdon North in the 1950s. The signal had recently replaced an earlier design. The adjacent goods sidings and shed still handled a considerable amount of traffic. (8.4.56.)

Chapter Two

MOTIVE POWER AND ANCILLARY EQUIPMENT

Locomotive depots in the days of steam were fascinating places. The larger sheds often had repair facilities, where space was usually at a premium. This frequently resulted in a partly dismantled engine being moved outside to await spares or further attention, and strange sights were not unusual. There were engines without wheels and pony trucks and bogies with their valve gear dismantled. Parts were normally left lying on the top of the running plate, which was also a favourite place to put coupling rods. On numerous occasions engines could be seen with various parts supported on wooden blocks. Pictures of locomotives under repair are included in this section.

Working outside on an engine in the depths of winter must have been far from pleasant, especially when handling cold metal parts, to say nothing of what the elements could throw at the fitters. Repairs carried out in the shed itself could also present problems: lighting was often poor, space cramped and, in the majority of cases, there was no heating (unless the engine itself was still in light steam or in the process of cooling down).

Maintenance of steam locomotives required an army of staff in addition to the fitters, for tasks such as boiler washouts and the removal of ash and clinker – one of the dirtiest jobs. Staff were also needed for coaling engines, and maintaining and issuing spares, oil, rag, etc. In addition, especially at larger depots, a number of staff were required to work out engine

This small turntable at the rear of Huntingdon shed saw little use. The LMR locomotive working a weekday only pick up goods from Kettering to Godmanchester was the only one that used it daily. (23.10.54.)

On Sunday mornings during the 1950s it was not uncommon to see a London-based tank engine running light to Doncaster Works, some of them sadly never to return, as diesels took over their duties. Here, J52 no. 68819 heads north having taken on water at Huntingdon. This veteran, built in 1899, was to remain in service for another year, being withdrawn in June 1956. (7.2.55.)

and staff rosters and to maintain the necessary records. In the final years of steam, locomotive cleaners became scarce although regular cleaning of engines was still undertaken during the 1950s. At one time this job was the first step to becoming an engineman. Towards the end of steam many engines returned from a general overhaul in immaculate condition and with a new coat of paint; in most instances they slowly became covered in an accumulation of grime and oil, and were never cleaned. Goods and shunting engines were often so dirty it was difficult to read their numbers or to make out the British Railways emblem. Fortunately this did not apply to those locomotives working on passenger services, which were normally kept in good external condition, especially so in the case of the A4s allocated to King's Cross, which were kept in immaculate condition right to the end of their service.

Enginemen were also required for locomotive disposal after they came on shed; this involved coaling, watering, dropping fires and moving them into the right position for their next duty, or for maintenance and repairs. There was a constant movement of engines, especially at larger sheds. This did of course have one advantage for enthusiasts, in that while all this was going on it enabled pictures to be taken of engines that were kept tucked away at the back of a shed as they were towed outside for a short period of time.

Interesting items of rolling stock were often to be seen at motive power depots, although most enthusiasts would not give them a second glance. Wagons for sand and oil, and old tenders often enjoyed a second lease of life as sludge-carriers where water softening plants existed, or as snowploughs. For most of the year the latter were simply dumped in a siding. Many of the principal sheds of a district had a breakdown train consisting of steam cranes of varying sizes. The make-up usually included old coaches modified for use as tool vans and mess facilities.

The North Eastern and Eastern Region did not issue permits to individuals to visit their motive power depots, so it was necessary to join an organised visit. You could try to get permission to visit from the Locomotive Foreman but this was often refused, and a few devotees even turned up regardless, perhaps hoping not to be detected.

Doncaster Works, with the frame of J6 no. 642203 nearest the camera. The frames and cabs of both engines are mounted on trolleys for easy movement. Visits to locomotive works were often available on one day per week or as part of organised visits by railway societies. (7.11.54.)

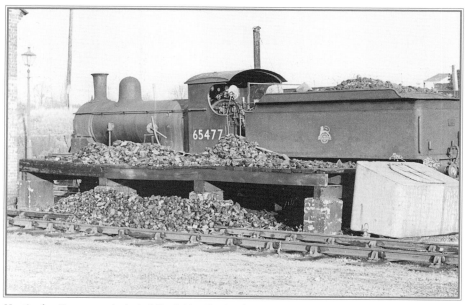

Huntingdon East was a sub-shed of Cambridge. During the 1950s a J15 class 0–6–0 was outstationed here on a ten-day rota. Behind the shed was a small turntable, while coaling facilities were a simple staging and water was from a hosepipe. Only on very rare occasions did an engine use the shed itself, and they were normally to be seen standing alongside the coaling stage. (9.5.54.)

The works plates on ex-Great Northern Railway passenger locomotives were brass, while those on goods and shunting engines were usually cast-iron. This is an example of the latter, fitted to J52 no. 68810, built in 1897 and withdrawn in November 1955.

This snowplough, no. 973, allocated to March depot, had been converted from a much larger locomotive than those found at other depots. The snowplough, just visible, was mounted below the buffers. (24.1.65.)

The locomotive servicing point enabled engines to take on coal and water before proceeding to the shed; it was also used as a holding point for engines ready to work trains out of King's Cross. A4 no. 60009 *Union of South Africa* had just worked in with 'The Elizabethan'. (9.61.)

Several A3 class Pacifics ended their days at New England. One of these was no. 60054 *Prince of Wales*, seen here near the King's Cross servicing depot in a very run-down state – note the burnt smokebox door. Many trains were already diesel-worked at this time. (9.61.)

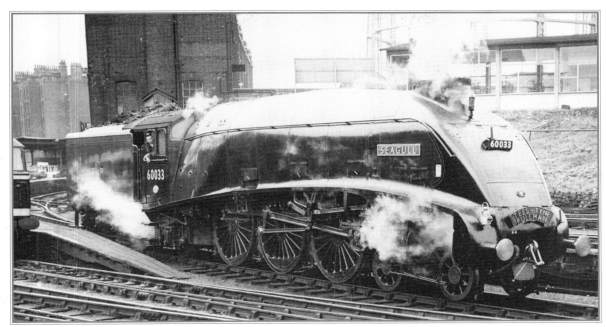

King's Cross was generally known as 'Top Shed', and its A4s were generally maintained in good external condition right up to the end of steam. Here, no. 60033 *Seagull* had just arrived with the 'Tees–Tyne Pullman'. (9.61.)

J50 no. 68991 was the last of the class to be built. Completed in August 1939, it was withdrawn twenty years later, and is seen here taking on water at Huntingdon on its final journey to Doncaster. Sunday mornings were a favourite time to send engines on their last trip. (8.61.)

Only a handful of V2s received double chimneys. This is no. 60880, pictured while on 'standby duty' at the north end of Peterborough station. The 'standby' locomotive was frequently called upon in the early 1960s to take over from ailing diesels. (6.9.62.)

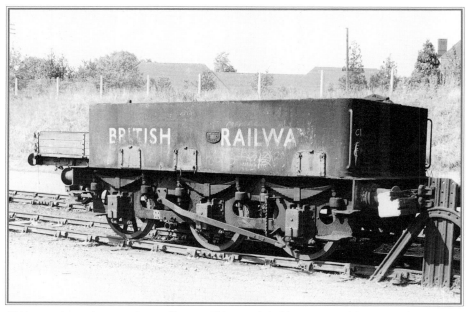

Old locomotive tenders were always of interest. This example had been converted for use as a sludge-carrier. It was still carrying British Railways lettering from its steam days, and a plate had been fixed near the letter 'R' with the number 981707.

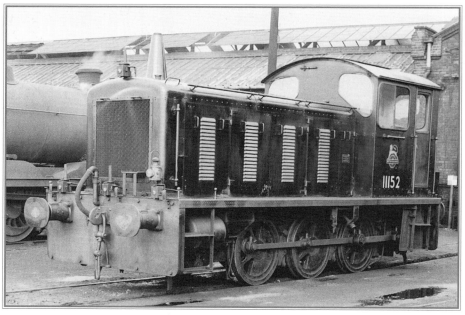

Most visitors to Immingham motive power depot would have recorded the number 11152 and quickly moved on. At this time the diesels were regarded as being little threat to steam in the foreseeable future. (25.8.57.)

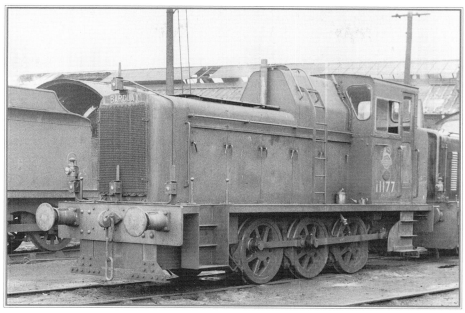

Immingham depot still had a large allocation of steam when this picture was taken, although several diesel shunters were already on the depot's books. One of these was Barclay no. 11177, seen here standing alongside an 04 class 2–8–0. (25.8.57.)

Repairs under way on N5 0–6–2T no. 69283 and a J6 0–6–0 at Retford Great Central depot. Note the lifting equipment above the J6, which has a 'Not to be moved' notice above the buffer beam. (25.8.57.)

The massive water-crane situated at the north end of Cambridge station was in frequent use. Here the enginemen of Britannia class no. 70002 take a well-earned break as the tender fills with water. (10.10.55.)

For many years F6 2–4–2T and C12 4–4–2T class locomotives were used on passenger pilot duties at Cambridge. There was a surplus of engines during the early 1950s and several were placed in store alongside the depot, including two of each class. F6 no. 67227 was in good external condition and was to see further service; it went on to become one of the last survivors of the class, being withdrawn from Colchester in May 1958. (14.4.52.)

Oil, grease, water and ash made working at the ash pits a very unpleasant job. Here, J67 no. 68609 has just completed its work for the day and the fire is being cleaned out at Cambridge motive power depot. (14.11.51.)

Two C12 class 4–4–2Ts were stored behind Cambridge depot for some time in 1952, before eventually returning to service. Both had their chimneys covered with tarpaulin in the usual manner. This class was introduced by the Great Northern Railway to work suburban services from King's Cross. (14.4.52.)

Only on very rare occasions was the J15 0–6–0 on Huntingdon pilot duty steamed for engineering works. On most weekends this was the scene at the small sub-shed. No. 65477, one of the regular engines on this duty, presents a picture full of useful information for modellers. (17.11.51.)

The names allocated to the Sandringhams of B17 class and later rebuilds to B2 class were a mixture of country houses, football teams and club colours. There were a few exceptions, with two East Anglian Regiments, *City of London*, and the 'royal' engine, *Royal Sovereign*. No. 61603 *Framlingham* was named after a property near Saxmundham.

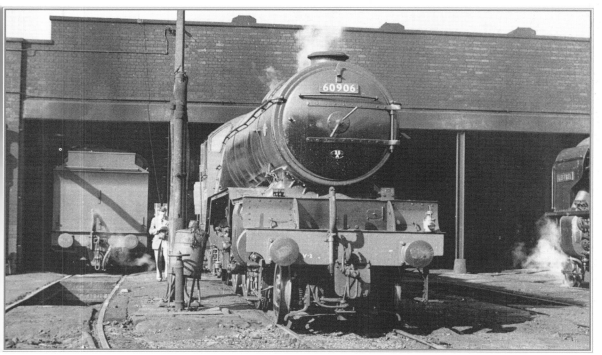

New England V2 no. 60906, fresh from works overhaul, stands outside Copley Hill shed with one of the depot's allocation of A1 class Pacifics alongside. This depot transferred to the North Eastern Region in 1956. (13.5.56.)

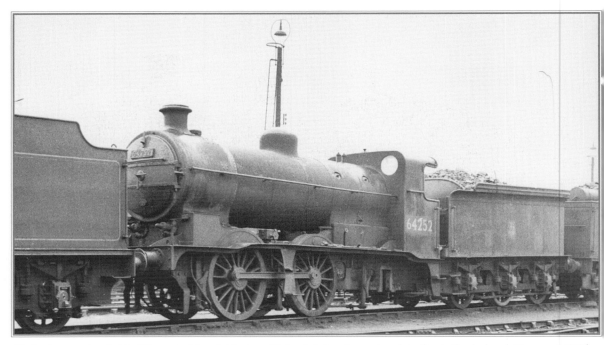

On the few occasions I visited Doncaster I found at least one J6 class 0–6–0 under repair and standing in the shed yard. Here, no. 64252 has been temporarily reduced to an 0–4–0: quite what the effect would be on the frames is anyone's guess! (24.6.56.)

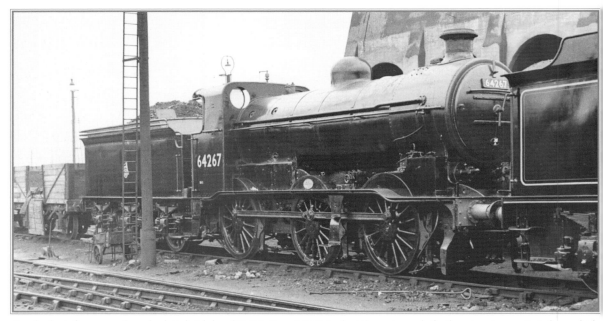

Resplendent after a general overhaul, Colwick-based J6 0–6–0 no. 64267 stands at Doncaster ready for the return journey to its home shed. The J6s carried brass workplates on the middle splasher. (24.6.56.)

Examples of the N1 class 0–6–2Ts had a long association with Ardsley depot, although they were in fact principally designed for work in the London area. No. 69481, built in 1912, was within days of withdrawal when this picture was taken at Ardsley. (13.5.56.)

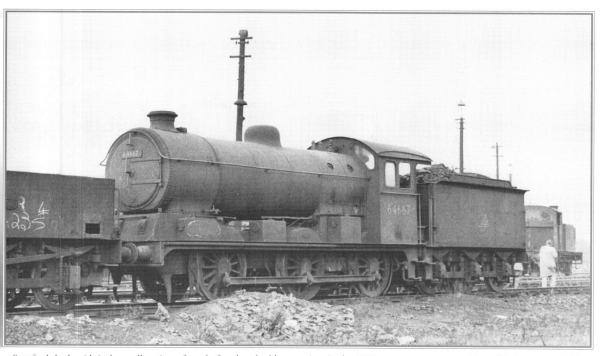

Stratford shed, with its huge allocation of nearly four hundred locomotives in the 1950s, covered a vast area. Here, J19 0–6–0 no. 64667 stands alongside a converted tender. The visits to this shed were never long enough, as it took some time just to cover the yards! (29.10.55.)

Locomotives in varying stages of repair were normally to be seen at Stratford shed, as was the case with B17 no. 61667 *Bradford* of Colchester depot. On occasions there was a very smoky atmosphere owing to the large number of engines present. (29.10.55.)

Sandringham class no. 61632 *Belvoir Castle*, seen here at Stratford, had an interesting history. Built as a B17 it was rebuilt as B2 in July 1946, running with a tender from one of the Gresley P1 class 2–8–2s. In October 1958 it was renamed *Royal Sovereign* following the withdrawal of no. 61671, but this was short-lived as it was condemned in February 1959. (13.11.55.)

Work-stained and under repair, J15 0–6–0 no. 65443 presented a sorry sight at Stratford. The dome had been removed and temporarily placed over the safety valves. Note the burnt smokebox door. Despite its grimy condition, this engine was to remain in service for four more years. (13.11.55.)

An engine that I had seen and photographed at Cambridge on numerous occasions was J66 no. 68383. When this picture was taken it had made its last journey from Staveley to Stratford, where it was built in 1888. It was the last of the class in running stock, and was withdrawn in October 1955. (13.11.55.)

B1 class no. 61267 stands ready for its next duty at Bradford Hammerton Street shed. Several B1s were allocated to the depot during the 1950s, but it was an early closure to steam, in January 1958. (13.5.56.)

Resplendent after a general overhaul at Stratford Works, this is J17 no. 65514 ready to return to its home depot, Norwich. In the background are the stored 4–4–2T no. 41976 and an F6 2–4–2T. (29.10.55.)

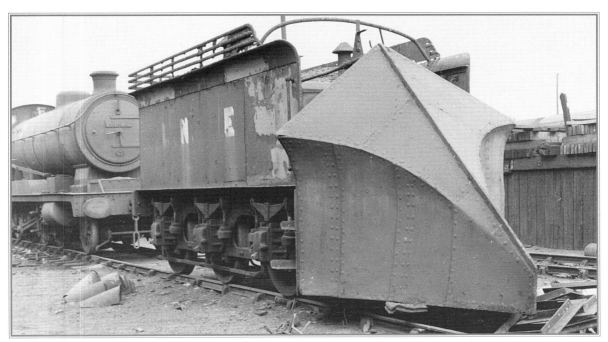

This old tender, renumbered no. 951580, photographed at Mexborough, had been fitted with a sizeable snowplough. Quite what would have happened if it had run at speed into a deep snowdrift is difficult to imagine. The old lettering LNER and later NE was easily visible. (24.6.56.)

Visits to Doncaster shed were always eagerly anticipated, wondering which Pacifics would be present, either awaiting or ex-works. These could be locomotives never seen in the London area, with the exception of the 'Top Link' A4s on the non-stop summer workings. A3 no. 60043 *Brown Jack* is seen here awaiting a general overhaul. (23.9.56.)

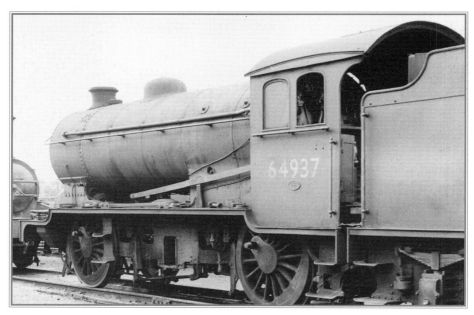

Locomotives under repair because of a 'hot box' were by no means uncommon during steam days. These were frequently moved out into the shed yard to await the return of their bearings. This was the case with J39 no. 64937, photographed at Lincoln. (25.8.57.)

In 1957 diesel shunting locomotives were starting to take over duties previously worked by steam. As a result engines were often placed in store, and this was the case at Immingham. J94 0–6–0ST no. 68074 was one of the batch built by Andrew Barclay & Co. It was eventually returned to traffic and was withdrawn in October 1962. (25.8.57.)

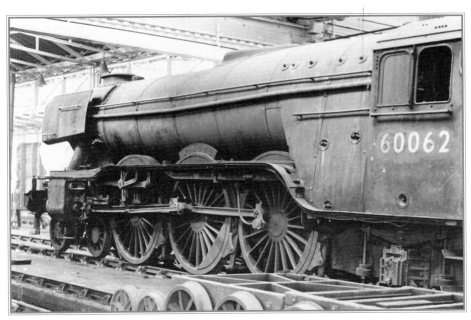

Almost certainly the last locomotive to visit the repair shops at New England was A3 no. 60062 *Minoru* in December 1964. Less than one month later the depot closed. Several A3s ended their days at New England, going for scrap when the depot closed. (6.12.64.)

A3 class no. 60106 *Flying Fox* had been well turned out to work the 'London North Eastern Flier', seen here taking on water at Peterborough. New England engines worked local services to King's Cross at this time, along with the other A3s. (2.6.64.)

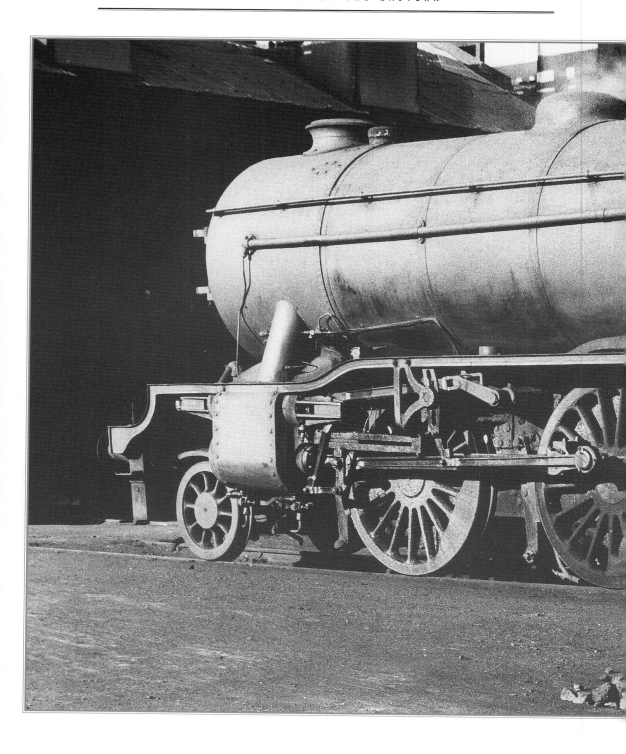

Several March K3 class 2–6–0s, including no. 61880, pictured at rest in the yard just where they had ended their final duty. The remaining coal has been thrown out of the tender. The steam just visible over the engine's dome came from a B1 standing alongside which was still in service. March had a considerable number of withdrawn and stored engines present in the early 1960s. (9.9.62.)

A1 class no. 60128 *Bongrace*, photographed at New England. The overhead locomotive water gantry in the background supplied water to nine tracks. This A1 would probably have worked in from York. (6.12.64.)

New England depot had an overhead system for locomotive water supplies. This depot was in its last weeks of operation when I took this picture of a grimy standard 9F 2–10–0, no. 92181. Note the clinker and ash lying around, a very different scene from the heydays of steam. (6.12.64.)

Old locomotive tenders were used for many purposes, such as snowploughs. This example, lettered 'Snowplough no. 955', was pictured at New England just prior to the depot's closure. It had been repainted black, but exposure to the elements resulted in the British Railways lettering still being visible. (6.12.64.)

Known locally as the 'Wedge', the massive coaling tower at March depot could be seen from many miles away, standing out in the flat Fenland landscape. Taking such a picture in the height of steam days would have been impossible, as the sidings would have been packed with engines. (9.9.62.)

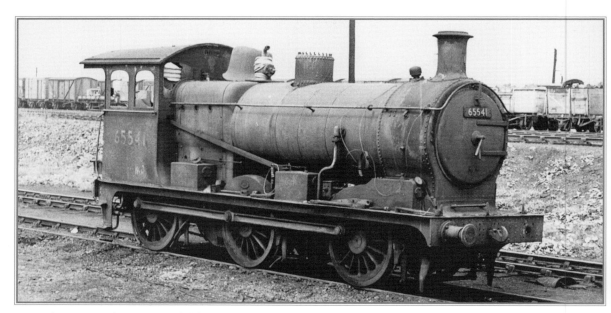

Over the years many locomotives ended their days as stationary boilers at various sites. At steam depots they supplied hot water for boiler wash outs. J17 no. 65541 was withdrawn from service stock in September 1962 and spent a comparatively short period as a stationary boiler. When this picture was taken it had been towed out into March yard ready for scrapping. The dome cover was lying on the top of the boiler, and one buffer and coupling were missing. (14.7.63.)

On withdrawal from service stock several B1 class 4–6–0s were transferred to departmental duties and used for carriage heating. Departmental locomotive no. 21, pictured at March, was formerly no. 61233. The couplings were removed from these engines to prevent them being used for any other work. (21.6.54.)

This steam crane was used at New England depot to handle the vast quantities of ash produced by the steam locomotives. When this picture was taken the depot was just one month from closure and the crane had been dumped in a siding. (6.12.64.)

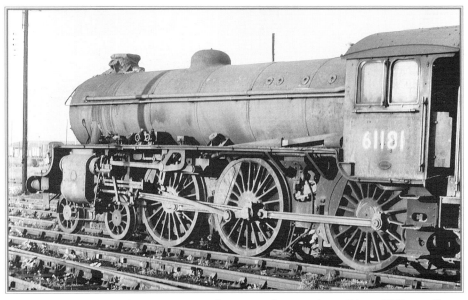

During the early 1960s locomotives were to be seen stored at many depots. B1 no. 61181 was still serviceable when this picture was taken at March. The chimney has been covered in the time-honoured method, with tarpaulin tied round the chimney. (9.63.)

Lincoln St Marks was a sub-shed of the main depot, shedcode 40A, from 1953, and also supplied the motive power for cross-country services. J39 no. 64714 stands in steam awaiting its next duty. Note the bent and buckled fire irons in the foreground. (14.8.55.)

The majority of turntables were hand-operated, as shown here at Bury St Edmunds, where the enginemen were busily engaged in turning D16 4–4–0 no. 62543. It could be heavy work if the engine was not positioned correctly when it was ready for turning. (2.7.55.)

Numerous older locomotives had little in the way of protection for the enginemen, especially when working tender first. To counter this, a sheet was frequently suspended from the rail that extended across the tender (the end of this rail can be seen above the handle of the fire iron). The letters RAI refer to the route availability, but J15 class 0–6–0s were permitted to go almost anywhere. (11.7.53.)

The scrap lines at the Eastern Region works were often full of veteran locomotives, their working days over. This is J70 tram loco no. 68225, standing at Stratford Works next to a withdrawn F4 class 2–4–2T. The J70 must have been towed on its final journey as the coupling rods and guards had been removed and part of the motion tied up. Note the lamp had been left on the engine. No. 68225 was officially withdrawn in March 1955. (7.5.55.)

N7 tank no. 69727, fresh from overhaul, at its home shed, Stratford. It would not have remained in this condition for very long at this massive, often smoky depot, which was home to a great many more of its classmates (used principally on Liverpool Street services). (7.5.55.)

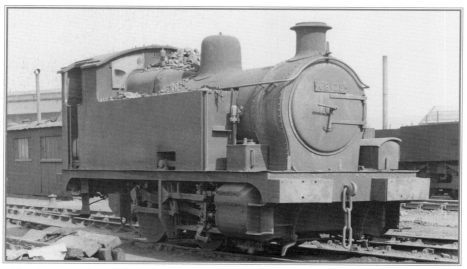

The pioneer Y4 class 0–4–0T no. 68125 had just four months left in service when this picture was taken. Five examples were built, and these sturdy, powerful engines looked much younger than they actually were. Note the solid buffers and coal piled on the firebox top. The last example in service stock outlived its classmates by several years at Stratford Works. (7.5.55.)

Coaled, watered and ready for its next duty, this is Sandringham no. 61613 *Woodbastwick Hall* pictured at Stratford. Sandringhams could be rough riding as a general overhaul became due. (7.5.55.)

There was little protection for enginemen on the ex-Great Eastern J15 class 0–6–0s, especially when working tender first. The locomotive controls can be clearly seen in this picture. The J15s were principal goods locomotives in their early days and despite their comparatively small size they were capable of handling large loads. (1952.)

This giant set of sheer legs was once in regular use at Yarmouth South Town shed. D16/3 class no. 62613 was officially allocated to Vauxhall depot, which served as a sub-shed of Yarmouth from 1957 until both closed in 1959. Note the old coach bodies in use as stores. (20.8.57.)

A3 class no. 60071 *Tranquil* was among the last members of the class to remain in service, being withdrawn in October 1964. Built in 1924, it achieved an amazing mileage figure, covering 2,250,000 miles during its forty years' service. Note the LNER works plate, the centre of which has the small insert '71' bolted on.

Yarmouth Beach was a Midland and Great Northern depot. The shed roof had long since gone when this picture of J17 no. 65586 was taken. Note the single line tablet exchange apparatus fitted on the tender. The depot closed in February 1959. (18.8.57.)

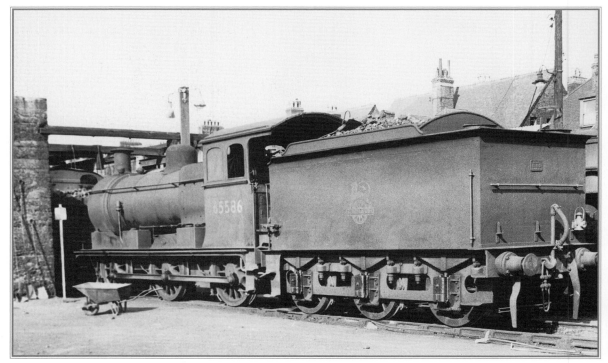

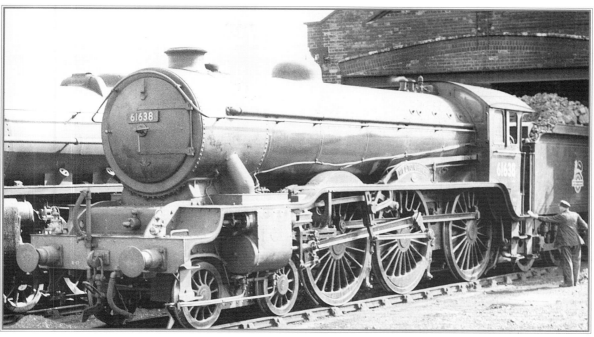

Sandringham no. 61638 *Melton Hall* of March depot pictured at Yarmouth Vauxhall shed; on the adjacent track is another visitor, a class 5. In the 1950s there were three depots at Yarmouth, two of Great Eastern origin and one of Midland and Great Northern. (20.8.57.)

The J52 class was another Great Northern Railway workhorse. These saddle tanks were once to be seen operating from many depots, with quite a number in the London area. No. 68882 was in a deplorable state, with the emblem and number barely visible, when under repair at Doncaster shed. Part of the coupling rods had been removed and can be seen lying on the running plate. (23.9.56.)

Another J6 0–6–0 in trouble with bearings. No. 64179 had been towed out to the shed yard at Doncaster. There is little to support the rear end of the engine as the third set of driving wheels has been removed. J6s seemed to be especially prone to this trouble, and over the years I photographed a number of examples undergoing repairs. (23.9.56.)

A3 no. 60037 *Hyperion* was one of a batch completed in 1934. When this picture was taken at Doncaster shed it had just been through the works for a general overhaul. It was withdrawn from service in December 1963. (23.9.56.)

First introduced by the Great Northern Railway in 1911 to the design of H.A. Ivatt, the J6 0–6–0s were very sturdy and useful. No. 64256 was completed in April 1919. It had just received a general overhaul at Doncaster when this picture was taken – almost certainly its last as it was withdrawn in April 1960. (24.6.56.)

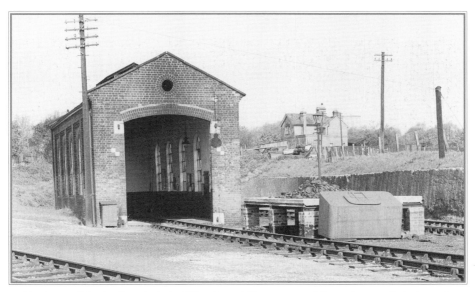

This picture of Huntingdon sub-shed will be of especial interest to modellers as it was typical of a number during the 1950s. In the background can be seen the stationmaster's house; this and the shed building have long since been demolished and the site where they once stood is now part of the A14. (9.5.54.)

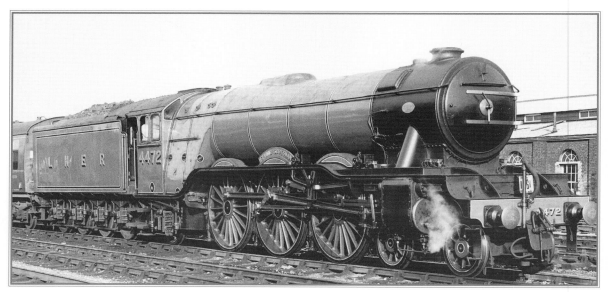

The legendary (restored) 'Flying Scotsman' pictured at Cambridge with its support coach. The locomotive was purchased on withdrawal from service in January 1963. At the time this picture was taken it had been repainted in LNER green-lined passenger livery and was used for privately chartered passenger trains. Naturally it attracted considerable attention during its short stop. (1.10.64.)

Lime from a water-softening plant heavily stains the side of this old tender. Still lettered LNER, it was at one time coupled to ex-Great Central D6 class 4–4–0 no. 5871. Quite a number of old tenders from withdrawn engines enjoyed a second lease of life on this work. (25.11.55.)

Chapter Three

ROLLING STOCK

Long since disappeared from our rail system are those seemingly endless slow, loose-coupled goods trains, clanking and rattling their way to distant destinations. The wagons produced their distinctive sound as they buffered up in response to any speed or gradient variations.

Countless thousands of wagons, in a huge range of types and sizes, were to be seen in steam days. They ranged from standard types introduced by British Railways to ageing wooden-bodied examples, many of which were owned at one time by private companies. These were the days when goods were extensively transported by rail, and coal and mineral traffic, perishables, fruit, fish and other items were commonplace. There were special wagons for some loads, such as gunpowder, sand and china clay – some of which are still moved today by rail in bulk loads.

'Well' wagons were commonly used to move farm machinery and other bulky loads, while steel girders and wood were transported on long bogie flat bolster trucks. Steel tube was normally loaded on to long-sided wagons. Bricks had a special type of wagon for their transport, which comprised a long-sided truck with two four-wheel bogies.

Every goods train in steam days had a brake van. The job of the goods guard must have been rather lonely, especially at the end of a long mixed goods or coal train. These could, if they lost their path, spend lengthy periods waiting in sidings. In complete contrast, the guard on a fast goods could find it very lively, especially if he was unfortunate enough to have a rough riding brake.

With all the individual loads, vast marshalling yards were necessary to sort out the wagons for the next section of their journey. One such yard was the huge complex at Whitemoor, March. Here a constant stream of goods wagons were received, sorted and dispatched. Whitemoor was a hump yard, where wagons were propelled upwards and then travelled by gravity into the numerous sidings. All this has passed into history.

During the 1950s you could find numerous veteran coaches still in use, many on branch lines and cross-country routes. Principal express trains often included examples of Gresley-designed stock. At the other end of the scale, old coaches, their normal service days long over, were to be found on engineers' trains and other departmental duties. Some were six-wheelers dating back to Great Northern days.

Vast numbers of wagons and coaches ended up in scrapyards; the wood was usually burnt and the metal recovered.

This 20-ton brake van, no. E235176, was typical of hundreds in regular use. Note the observation position with its small windows and the chimney coupled to the small stove, which provided heat and limited cooking facilities.

This 16-ton steel-bodied mineral wagon, no. B92146, has just been emptied. Coal spilt during the process still remains to be cleared up and a typical coal-sack hangs over the side of the wagon.

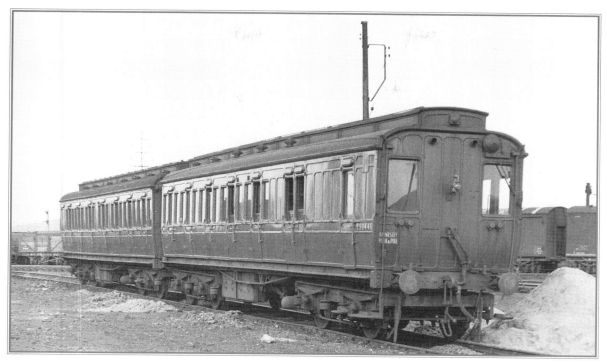

A two-coach push-and-pull set was used to convey staff to and from Annesley motive power depot. The leading coach is E5344E. Note the clerestory roof coaches and driver compartment nearest the camera.

This fitted Algeco tank wagon, no. 501804, was built by Metro Cammell. It was used to convey diesel oil.

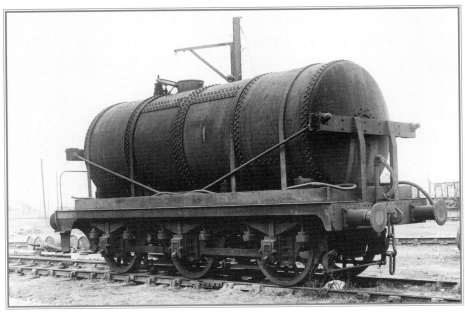

Another interesting example discovered on one of my visits to March was this six-wheeled tank wagon, DE962033, which carried no identification other than a barely visible stencilled-on number.

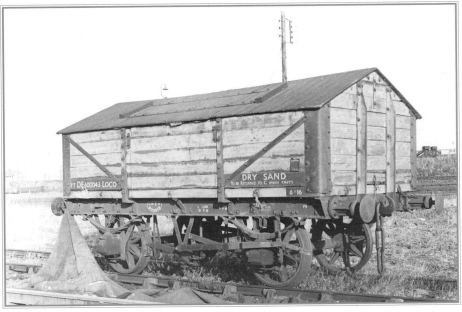

This interesting wagon was also discovered in a siding near March motive power depot. It was used to transport dry sand for locomotive use. The plate reads LNER 60043, and the lettering on the side reads '7T D600043 Loco' and 'to be returned to C when empty'. It would appear to have been covered at one time with the tarpaulin lying on the tracks.

The lettering on ore wagon B433871 was almost completely obliterated by rust. It was typical of the state of many in the early 1960s.

Tank wagons similar to this one operated by the Esso Petroleum Company Limited were a familiar sight. The wagon was fitted to run in vacuum-braked trains. Oil had run down the tank side at either unloading or loading, and can be clearly seen under the ladder.

No. M360532 was typical of the vast number of five-plank wagons still in service during the 1950s and 1960s. There was no trace of the old lettering on the side so the wood may well have been replaced at some stage.

Long-wheelbase five-plank wagons were used for transporting tubes. B732104 was fitted with roller bearings and may have been working from the tube works at Corby.

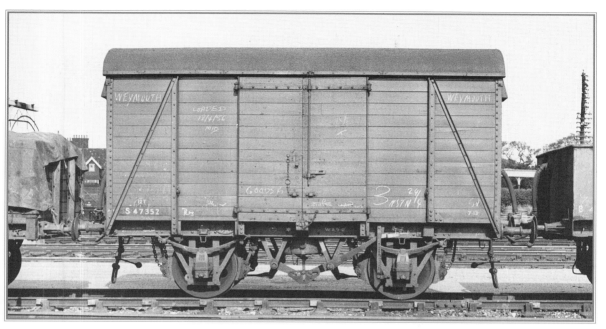

Goods vans travelled far and wide on British Railways. This 12-ton fitted example, no. S47352, carries several chalked inscriptions, among them Weymouth, easily seen on this broadside view.

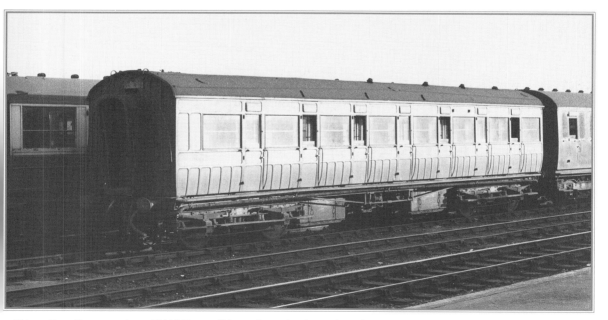

This ex-Great Eastern coach no. E61497E was still in regular service during the early 1950s. Only a few enthusiasts paid any attention to coaches at this time. Cambridge, where this picture was taken, was very interesting from the locomotive aspect.

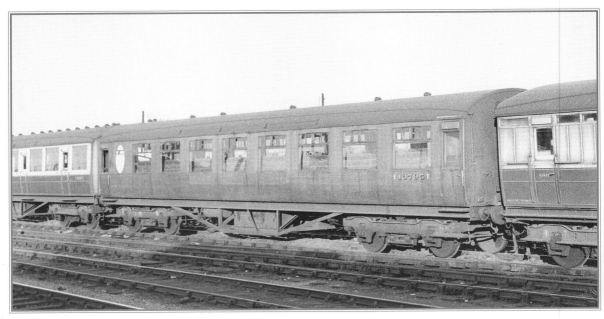

Resplendent in teak finish, this 3rd class open coach no. 13795 had had the letter E added at each end of its number. It was standing among a variety of other coaches in the sidings at Cambridge station when photographed.

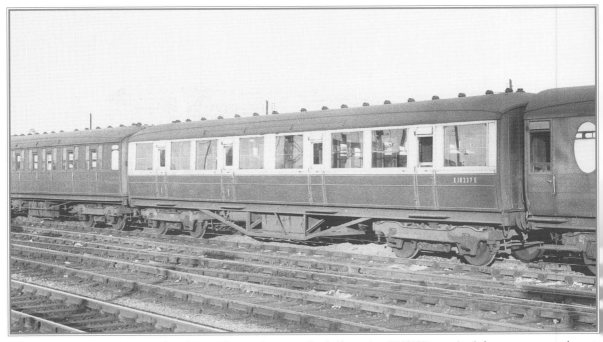

During the 1950s a wide selection of coaching stock was to be seen at Cambridge station. E18237E was a 1st–3rd compartment coach, next to a more recent introduction of LNER design.

During the 1950s the Chapman Chemical Co. Ltd operated a weed-killing train. Spraying was done using a converted brake van, no. 601014. The pipework and spray-heads can be seen in this picture, together with a number of the tank wagons used to convey the chemicals. Standing alongside is a similar vehicle bringing fresh supplies.

Another view of the converted brake van fitted with equipment for weed spraying, often seen during the 1950s. No. 601014 was part of a train operated by the Chapman Chemical Co. Ltd.

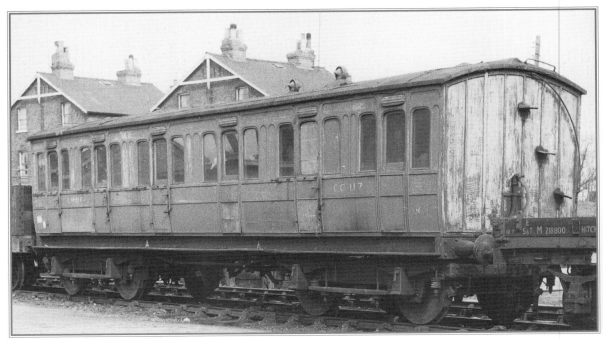

This veteran coach ended its days as a mess and store vehicle for the signal and telegraph department. Other than some internal modifications it remained very much as it had been when withdrawn from passenger service; it is still lettered LNER at the far end and has CC117 stencilled nearest the camera.

This massive 40-ton 'Walrus' bogie wagon, no. DB992498, was used for track ballast. Note the operating wheels and warning notice situated at the front bogie.

The 12-ton covered vans were widely used on all parts of British Railways. No. E181086 was an XP version, equipped to work in express trains. Alongside is a flat wagon, no. B460887, loaded with agricultural machinery.

This 12-ton van, no. W124905, was also equipped to work in express goods trains, as indicated by the 'XP' marked on the end nearest the camera.

This is a later version of an express goods 12-ton van. No. B784239 was built by British Railways and is of metal construction throughout, unlike many of those in service during the 1950s which had wooden bodies.

British Railways operated a door-to-door container service using Conflat A wagons. No. BD703413 was a typical example. Note the securing chains at each end of the container.

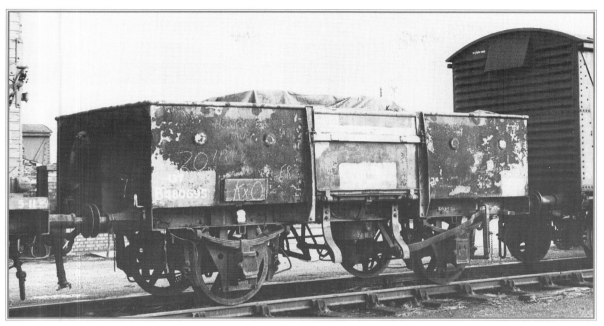

This 13-ton fitted steel wagon, no. B480695, pictured in poor external condition. This was not uncommon after their long exposure to the elements and the battering they sometimes received during loading.

Rough shunting and the many jolts experienced during travelling in goods trains could easily damage some loads. This 12-ton Shocvan, no. B851682, carried the XP marking, and was capable of working in fast goods.

Many thousands of 16-ton steel mineral wagons were built by British Railways. Seen here is no. B40139. Fortunately a number of these wagons have survived into preservation.

This 30-ton GUV bogie vehicle was used for transporting parcels, mail and many other goods. This example is no. M86964.

Typical of wagons that ended their days on service work, this is a 13-ton wooden-bodied wagon used by the engineering department. It was numbered DM283663. When this picture was taken it had come to the end of its working life, and had been stencilled 'Cond' – or condemned.

Low early evening light picks out clearly the details on this wooden-bodied wagon, no. DW117620, photographed in October 1966. It also had been condemned, and to prevent movement a clamp had been fitted to the wheel nearest the camera.

One feature that immediately catches the eye in this picture of 12-ton express goods van no. M205565 is the work panel at the end nearest the camera. Within a very short time this would have been rendered illegible.

This interesting six-wheeled tank wagon was discovered in a siding at March motive power depot. It seems to have been modified from an old locomotive tender chassis. The only identification was the number 961881 stencilled on the tank side.

The fruit season required considerable numbers of vans for a short period. At other times they were mostly to be found standing for months on end in sidings. No. B875255 is a typical example of a 12-ton XP ventilated fruit van.

During the 1950s older coaches were often seen standing in sidings, as they were only required for summertime excursions and specials. This 3rd class brake no. 62470E of Great Eastern origin, pictured in 1956, is a typical example.

Another interesting coach, this is ex-North Eastern E2945E, which spent some time in a siding prior to being used on summer excursion traffic.

These 12-ton loose-coupled vans were commonplace and this is no. B752452, pictured in 1956. Note the destination ticket held in the clip over the wheel nearest the camera.

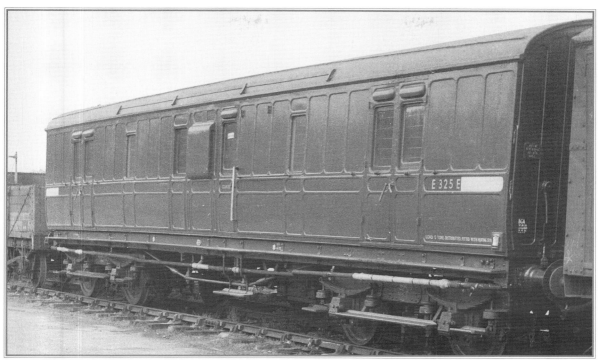

Bogie passenger brake no. E325E was built by the North British Railway in 1921. It was in good external condition when this picture was taken in 1955 and seems to have recently been repainted.

The transport of livestock by rail was still important during the 1950s. Nearest the camera is horsebox no. M42435; such wagons were capable of running in express trains and included a small compartment for the groom. (The door and window of this section can be seen at the far end.) Alongside is E70019, used for conveying racing pigeons.

Considerable numbers of fully fitted 12-ton vans were reserved for the transport of fruit only, and these spent long periods of time lying idle. This was the case with E226527.

Typical of many Esso tank wagons in the 1950s, no. 2351 is seen here in silver livery with large company lettering. This picture was taken at the Esso depot that was at Huntingdon during the 1950s.

Fresh from overhaul and resplendent in its shiny black livery, this is Esso 14-ton wagon no. 2301. These wagons were once a familiar sight on the railway.

Engineering department 25-ton slag wagon DE 470221. The lettering on the side reads 'Empty to Appleby slagheap, Stanton, Scunthorpe'. Note the side-hoppers that were operated by a member of the engineering staff for emptying the load while the wagon was moving along slowly.

Many goods could be easily damaged by rough shunting and the buffeting received while travelling in mixed goods trains. This is no. B722871, a fine example of the 12-ton 'Shock' wagon. These were fully fitted and could run in fast goods. The wagon is lettered 'Empty to Swansea (Eastern depot) WR'. A continual check was made on special wagons to ensure they were returned promptly to the originating depot.

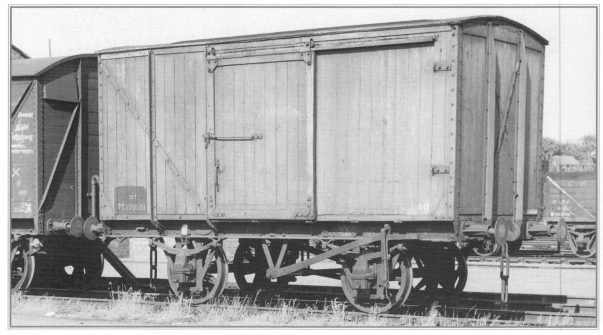

Compare this LMR 12-ton van no. M305033 to the one with E lettering (see page 99), and it is possible to see detail differences, such as that it is non-fitted (the small window in the end of the fitted example is not present), but the general construction is very similar.

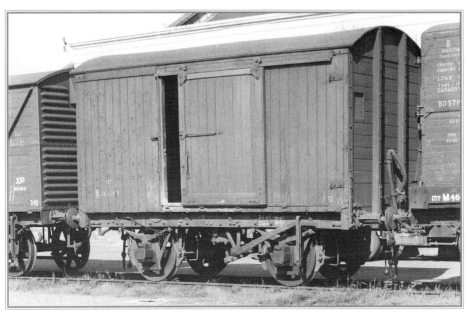

One type of wagon that was invariably used in pick up and mixed goods trains was the 12–ton fitted box type, similar to this one, no. E166140. Similar or slightly modified examples were reserved for fruit and fish trains.

This picture shows the springs, axle box and wheel of XP long–wheelbase van no. E1264. Note the LNER ZZ axle box cover.

The interior length of these XP fully fitted long-wheelbase vans was 39ft 1in, and their load capacity was 8 tons. This example is no. 1264.

These British Railways containers were a familiar sight in the 1950s. Here, 4-ton BD 5719B is seen ready to move on 13-ton flat wagon M460655.

This is the 14-ton Esso tank wagon, no. 1174, in the company's silver livery. Note the cables securing the tank to the chassis. These wagons were commonly seen during the 1950s.

These six-wheeled passenger brake vans were usually employed on parcel traffic. This is no. E2364E, photographed at Bury St Edmunds.

This old coach had been converted for use by the engineering department, and renumbered DE 320034. For some time it was to be seen standing outside one of the old steam sheds at March. In the background is one of the B1 class 4–6–0s converted to departmental use for carriage heating.

Huge numbers of these 'standard' mineral wagons were built. This example is no. B155069. These replaced some of the ageing wooden-bodied wagons.

Chapter Four

LINESIDE EQUIPMENT

Travelling by rail today is very different from fifty years ago, in steam days. On most lines the once-familiar signal-boxes, which existed every few miles, have gone completely. In most cases they have been replaced by colour light signalling and power boxes.

Signals varied widely, from a single arm on a post to huge gantries with numerous arms. The approaches to large busy stations or rail centres had an array of signals, many of which would have had a fogman's hut alongside, especially in the early 1950s, when dense 'pea soup' fogs were a regular occurrence in large cities.

Only a very few wayside stations were not equipped with a goods shed and the equipment required to handle goods traffic. Steam locomotives on the East Coast main line had the advantage of being able to take water at speed from the troughs positioned at certain locations. Passengers may well have realised what was happening as water sprayed the carriage windows, or perhaps they caught a glimpse of the necessary water installations at the trackside.

Another feature on the main line was the Travelling Post Office equipment, although I suspect few people understood its purpose even if they noticed it as they flashed by. While TPOs still run, the old system of dispatching and receiving mail en route without stopping has long since finished.

Over the years countless miles of redundant track have been ripped up and large areas of former railway land sold and redeveloped as part of the process of streamlining the railway, especially in cities and towns, leaving only the basic requirements.

Apart from controlling the main line, Huntingdon no. 1 signal-box was also responsible for the Cambridge–Kettering line. The latter was a single line operated by a token. The wooden pier for the token exchange can be seen leading off the main steps. (9.10.54.)

These two signal gantries were situated at the southern end of Huntingdon station. The two posts without arms once controlled access off the main line to Huntingdon East but they were taken out of service shortly after the war ended. (9.10.54.)

The fogman's hut was once a familiar sight, and platelayers were often called out at short notice to assist drivers in dense fog during the early 1950s. Note the set of repeater arms that were visible even if the main signals were not. (7.2.55.)

On the Down side at Huntingdon the Travelling Post Office lineside equipment consisted of the strong receiving net and four separate sets of arms, each capable of taking two pouches. When the train was imminent, the net would be swung into position and locked. The heavy pouches dispatched by the train ended up in the net. Note the warning indicator board, and the lamp holder, just visible in the foreground.

On the Up side at Huntingdon, the TPO equipment consisted of a receiving net and a single set of dispatching arms capable of handling three pouches. The North Eastern TPO used this in the early hours. Considerable quantities of mail were received but only late-posted mail was dispatched.

Trains approaching Huntingdon could continue on the main line or divert to the slow line. This was indicated by the two signals mid-way up the post, which worked in conjunction with the home signal.

During the 1950s you could often find interesting pieces of equipment in goods yards. An unusual example was this 1-ton capacity hand-operated crane mounted on a rather dilapidated wooden carrier; judging by the overall condition, this crane had seen little use for some time.

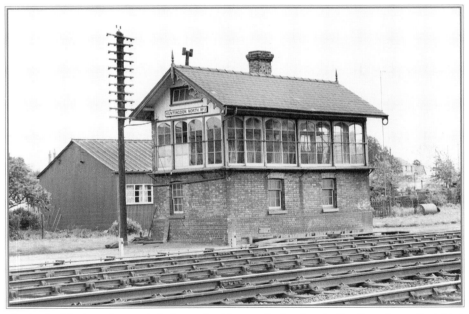

Huntingdon North no. 2 was typical of many signal-boxes to be found on the East Coast. It was situated just a few hundred yards from the northern end of Huntingdon station. (18.9.55.)

This fine group of signals was once to be found at the north end of Huntingdon station. Nearest the camera is the Down slow, with the signal controlling slow to main next. The last is the Down main. Note the repeater arms near the base. The signals have long since gone but the bridge now carries more traffic than ever, with the very busy A14 running over it on a new bridge. (25.8.52.)

A typical winter's scene on the East Coast main line. The bridge in this picture had received a new parapet at some stage in its long history. The crossing gate seen through the arch guards the LMR line to Huntingdon East. (4.1.53.)

This picture was taken near the small sub-shed at Huntingdon East and contains a wealth of detail. Note the ornamental gas lamp in the foreground and the warning sign. In the background is a typical platelayers' hut, as well as signals and associated equipment. (19.5.52.)

Above: Climbing signal posts to attend to the lamp is one railway duty long since gone on most parts of the system. The details of a typical 'home' signal can be clearly seen in this picture.

Left: One of the last examples of the Great Northern somersault signals. This splendid example stood on the main line south of Peterborough, at Paxton near St Neots. (20.6.53.)

The station at Abbots Ripton had long since disappeared when this picture was taken in 1966 but the signal-box can be seen in the distance with an impressive lattice post signal in the foreground.

Chapter Five

TRACK AND CIVIL ENGINEERING WORK

Modern methods of track laying have certainly speeded up the process of replacing worn rails; prior to the use of continuous welded rails, track was laid in sections, a process that required considerable gangs of men. As today, much of this work took place at weekends. For sizeable jobs two or more steam engines would be required to assist in the handling of track sections, as well as ballast transport and laying.

One locomotive type often used for these operations was the J6 0–6–0s allocated to a number of depots, often assisted by WD 2–8–0s. The Eastern Region also had a number of locomotives in departmental stock employed on various duties, for example at permanent way depots. One was Chesterton Junction Cambridge, where Sentinel Y3 class no. 42 was used for a number of years on shunting work. The engine made occasional visits to Cambridge depot, and when it needed major repairs it was dismantled on site. This Sentinel was condemned in July 1960. Other examples of the very similar class Y1 locomotives were to be found at permanent way yards at Peterborough and Lowestoft, as well as at the sleeper depot at Boston.

This wooden trestle bridge crossing the River Ouse between Huntingdon and Godmanchester was a problem for many years, imposing severe weight and speed restrictions. Originally double track, it was singled in the early 1950s and a 10mph speed restriction imposed. Where this bridge once stood there is now a very different, modern bridge carrying the A14 at high level. (20.2.54.)

Departmental locomotive no. 5 photographed at the small engineers' yard near Cambridge station. This Y3 was built by Sentinel in September 1930 and was taken into departmental stock in March 1953. It remained in service until November 1958. (8.11.54.)

This bridge carries the East Coast main line over the River Ouse just south of Huntingdon. The signal-box and home signal in the distance were temporary features while track alterations were carried out. Note the concrete platelayers' hut, typical of the period. (20.2.54.)

Speed restrictions were imposed after track re-laying had taken place. This example was in use on the East Coast main line near Huntingdon. I wonder how many pictures were taken of such items! (20.2.54.)

Steam travelling crane DE966215 was built by Grafton Cranes Ltd of Vulcan Works in Bedford. Note the rather battered metalwork. (11.4.65.)

This picture of steam crane DE906215 at New England provides useful information for modellers on the jib and the bogie 'jib runner', no. 320829. (11.4.65.)

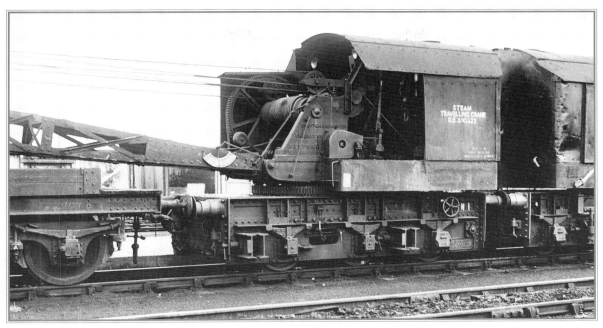

Steam travelling crane DE330222 was also built by Grafton Cranes Ltd of Vulcan Works. It is shown here in a siding at New England in company with a sister unit. (11.4.65.)

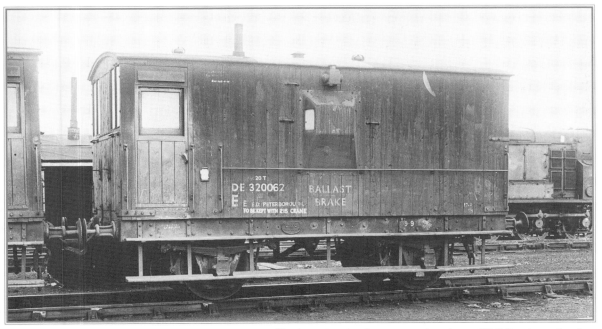

This unusual 20-ton ballast brake, no. DE320062, was operated by the engineering department and was lettered 'E D Peterborough to be kept with 215 crane'. (11.4.65.)

Grafton steam crane DE330222 and jib runner DE320683 await their next call to duty in a siding at New England. Much of the work carried out by these units took place during weekends when they could take complete possession of the line. (11.4.65.)

Track recovery had reached Godmanchester when this picture of J15 no. 65420 was taken. The locomotive was officially allocated to March at the time and was working from New England, involving a night and morning run on the main line. The J15 was one of only a few types that could cross the wooden trestle bridges because of the severe weight restrictions. (31.8.61.)

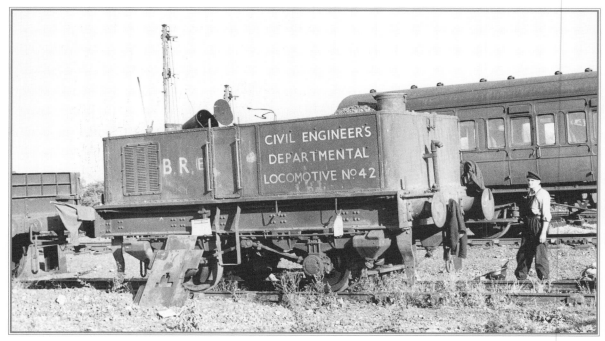

On a visit to Chesterton Junction permanent way depot I was amazed to find the resident departmental Sentinel Y3 no. 42 dismantled for repairs. The cab and engine have been removed, and the locomotive looks more like an old tender in this picture. (2.10.55.)

This is the Sentinel's sizeable engine unit, lying in the yard at Chesterton Junction permanent way depot while undergoing repairs. The cab is also seen in this picture. Unfortunately, I was so intrigued by the Sentinel that I failed to take a picture of the veteran coach in the background. (2.10.55.)

Track-laying equipment being moved into position by a diesel shunter. While this was a great improvement on previous methods, it was designed for use with rail laid in sections – a very different method from today's continuous welded rail. (2.10.55.)

J15 no. 65420 standing on one of the wooden trestle bridges that bedevilled the Huntingdon– St Ives section. In the background is the long since demolished Godmanchester Mill. Despite the problems they caused, the huge wooden timbers of these bridges did not yield easily when it came to removal. (31.8.61.)

This small platelayers' trolley was kept at St Ives and used to transport materials. In the background is a neat well-trimmed hedge. After the railway had closed and the track and buildings were cleared, part of this hedge survived for many years. (8.10.51.)

This picture shows not only the tracks but also details of points, rodding and signal wires, together with a lattice-type signal post typical of many once to be found on the East Coast main line.

In this picture of Abbots Ripton, taken in the mid-1960s, nothing remains to indicate that there was once a country station and signal box on the site, the only clue being the space between the tracks. Note the telegraph poles on both sides, at that time something to consider when taking pictures, wherever you were.

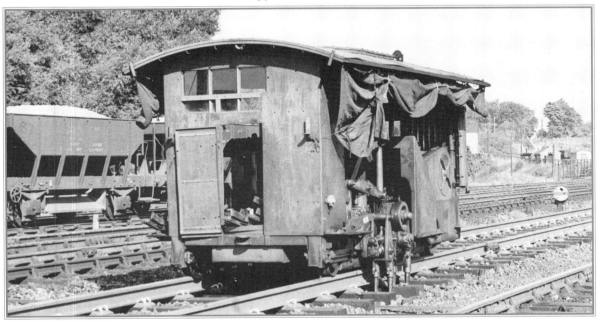

Looking rather like a garden shed on wheels, this Matisa ballast-tamping machine was busily working on the main line, which had recently been relaid with flat-bottomed rails. Note the tarpaulins hanging over the sides. (18.9.55.)

Engineering trains on the East Coast main line were usually worked by J6 class 0–6–0s and occasionally by WD 2–8–0s. Here, no. 64197 of Hitchin depot heads a train of lifted rail sections. This was a fairly easy (if perhaps boring) duty for the enginemen as it involved lengthy periods of inactivity. (2.10.55.)

Many old coaches ended their days on departmental duties. This six-wheeled example, DE 625550, was used as a ballast brake at Cambridge, where this picture was taken. (3.4.55.)

Chapter Six

MISCELLANY

Among the various commonplace items to be found on the railways were the cast–iron notices warning people not to trespass; many such notices dated back to pre-Grouping days. The railways must have cast them in huge numbers. Also in this section are a few pictures of accidents, where the huge 45-ton breakdown cranes can be seen in action. After the accident that occurred at Offord the main line was closed completely, in order to allow the huge amount of debris from the smashed wagons to be removed before the A1 class Pacific no. 60123 *H. A. Ivatt* could be lifted and re-railed.

While they are strictly outside the scope of this title, a very small number of early diesel shunting locomotives are also included, principally because they were taken in the mid-1950s and were to be found operating from steam sheds. At that time it is fair to say that few people, if any, fully appreciated just how quickly diesels would replace steam. Many railway photographers ignored them completely, and some of the early designs did not last very long so few photographs of them exist.

Compiling the pictures in this book has certainly taken me 'down memory lane', and I hope this book will bring back many pleasant memories to readers and will prove of interest to all modellers keen to include as much detail as possible.

The 45-ton New England steam crane pictured hard at work at Offord removing badly damaged wagons, many of which had been reduced almost to matchwood. This accident closed the East Coast main line for two days, resulting in lengthy diversions. Note the badly damaged van in the foreground. (8.9.62.)

Both the 45-ton breakdown cranes from Peterborough and Doncaster can be seen in action in this picture. On the extreme right A1 Pacific no. 60123 *H. A. Ivatt* is lying on its side, with the Doncaster crane preparing for lifting. (8.9.62.)

The Offord crash marked the end of the line for the badly damaged Pacific, which was the first A1 to be condemned. On the Sunday after the crash the engine was moved to a siding north of Offord crossing ready for its last journey to Doncaster. (9.9.62.)

The massive New England breakdown crane in action again, this time at Huntingdon East, re-railing the tender of J15 0–6–0 no. 65475, which had derailed on points while working a goods to St Ives. Lifting was in progress under the watchful eye of motive power depot staff. (3.8.55.)

Another picture of the New England crane as it prepares to lift a derailed van prior to starting on the J15. This powerful crane had a lifting capacity of 36 tons. To enable the crane to get into the correct position involved some careful shunting. (3.8.55.)

These yellow-painted articulated lorries and containers were a common sight during the 1960s, operating what was known by British Railways as a door-to-door service for containers. The container seen here is no. BD478298.

In steam days it was commonplace to see locomotives supported on blocks of wood in a shed yard while under repair. Barclay diesel D2953 , pictured on blocks at March, appears to have suffered bearing trouble. The top light had also been dismantled for attention. (6.12.64.)

For many years wagons at Yarmouth Docks were shunted by two double-ended Sentinel locomotives. No doubt the enginemen found the diesels much easier to operate, and doubtless preferred their more comfortable working conditions. No. 11103 is seen here complete with 'cowcatcher' and side-sheets. Note also the shed plate mounted on the end of the cab. (18.8.57.)

Trespass notices were common in the 1950s and many dated back to pre-Grouping days, as did this cast-iron Great Northern Railway example dated July 1896. The main body reads 'Persons trespassing upon any Railway belonging or leased to or worked by the Great Northern Railway Company solely or in conjunction with any other Company or Companies, are liable to a penalty of Forty Shillings under the Great Northern Railway Act 1896 and in accordance with the provisions of the said Act public warning is hereby given to all persons not to trespass upon the said Railways. By Order.' (1954.)

The railway companies must have produced thousands of these cast-iron warning notices. This is a London & North Eastern Railway example reading 'Beware of the trains, look both up and down the line before you cross'. Those that survive are now collectors' items. (1952.)

This cast-iron Great Eastern Railway trespass notice was much more to the point than the GNR example shown above. It seems that the plate had broken in half at some time and had been repaired. (1952.)